The Story of Life
EVOLUTION

BIG PICTURE PRESS

First published in the UK in 2017 by Big Picture Press,
an imprint of Bonnier Books UK,
4th Floor, Victoria House
Bloomsbury Square, London WC1B 4DA
Owned by Bonnier Books
Sveavägen 56, Stockholm, Sweden
www.bonnierbooks.co.uk

Illustration copyright © 2017 by Katie Scott
Text and design copyright © 2017 by Big Picture Press

7 9 10 8

All rights reserved

ISBN 978-1-78370-682-2
ISBN 978-1-78741-330-6 (eBook)

This book was typeset in Gill Sans and Mrs Green
The illustrations were created with pen
and ink and coloured digitally

Expert consultant: Dr John Rostron
Written by Fiona Munro and Ruth Symons
Designed by Mike Jolley and Wendy Bartlet
Edited by Ruth Symons

Printed in Latvia

The Story of Life
EVOLUTION

Illustrated by KATIE SCOTT
Written by FIONA MUNRO and RUTH SYMONS

BPP

6

Entrance
Evolution
Timeline of Life on Earth

13

Gallery 1
Precambrian
Precambrian Period; Cyanobacteria; Multicellular Animals

21

Gallery 2
Palaeozoic Era
Cambrian Explosion; Cambrian Hunters; Ordovician Period; Silurian Period; Devonian Period; Acanthostega; Woody Plants; Carboniferous Period; Tetrapods; Permian Swamps; Sauropsids and Synapsids; Mammal-like Reptiles

47

Gallery 3
Mesozoic Era
The Triassic; The Jurassic; Archaeopteryx, Cretaceous Period; Tyrannosaurus rex

59

Gallery 4
Cenozoic Era
Palaeogene Period; Ambulocetus; Cooling Earth; Neogene Period; Indricotherium; Megafauna; Smilodon; Man

77

Library
Index

THE STORY OF LIFE

Entrance

Planet Earth was formed about 4.5 billion years ago in a swirl of dust and rocks left over from the Big Bang. It was another 500 million years before the first life forms appeared. In fact, there were no plants or animals on Earth for around 90 per cent of its history.

When Earth first formed it was a hot, rocky planet, with no life at all. Violent volcanic eruptions created the atmosphere and oceans as the Earth gradually cooled. Since then, the planet has changed beyond recognition: continents have shifted; global temperatures have risen and dropped; and over time the right conditions for life have developed.

Life as we know it today is the product of millions of years of evolution. Our human ancestors first appeared around 200,000 years ago. This means that humans have been here for just 0.0004 per cent of the planet's history.

THE STORY OF LIFE

Evolution

Evolution is the scientific theory of how living things gradually change and develop over time to form new species. This process is largely driven by natural selection, whereby the organisms best suited to survive in their environment pass on their traits to the next generation.

The theory of evolution by natural selection was first developed by Charles Darwin and Alfred Russel Wallace in the mid-nineteenth century. It was based upon the observation that individuals within the same species show a wide range of physical traits, some of which are better suited to their environment than others. Individuals with a better chance of survival are more likely to reproduce, so they will pass their genes, and their traits, on to the next generation.

We now know that the theory of evolution links all life on Earth back to a single common ancestor in a process of change which is still taking place today …

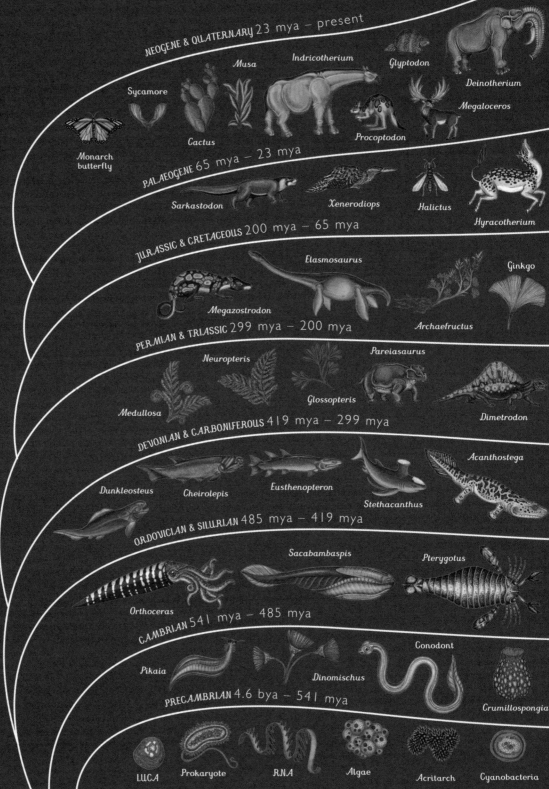

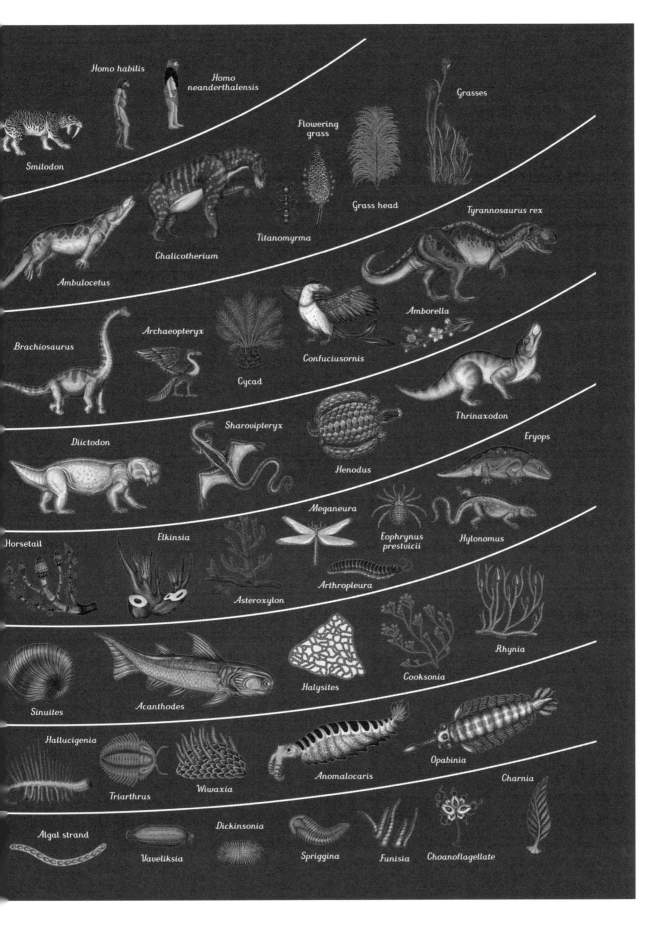

THE STORY OF LIFE

Gallery 1

Precambrian

Precambrian Period;
Cyanobacteria;
Multicellular Animals

PRECAMBRIAN
4.6 bya – 541 mya

Precambrian Period: First Life

The Precambrian is the name given to the first time period on Earth. During the billions of years of the Precambrian, the Earth formed and cooled. Volcanoes belched out gases, rocks formed from volcanic lava and the oceans condensed from atmospheric vapour. It is thought that life first appeared in these mineral rich waters as simple chemicals spewed through vents in the Earth's crust under the oceans, and reacted with one another to form more complex compounds.

Some of these molecules then combined and developed the ability to copy themselves, using the complex chemicals DNA, RNA and proteins — the building blocks of life. The next step was the protection of these chemicals with a membrane to form the first simple organisms.

The very earliest of these were single-celled organisms called prokaryotes — cells that do not have a nucleus (control centre) or any other subunits. Instead, all of their component chemicals float together, protected by the cell wall. It is thought that all life on Earth evolved from one such single cell, referred to as the Last Universal Common Ancestor (LUCA). This probably lived around 3.8 billion years ago.

--- *Key to plate* ---

1: **RNA (ribonucleic acid)**
Length: Less than 0.001 micrometres
RNA is present in all living cells and takes the form of a chain of molecules.

2: **Prokaryote**
Length: 0.1 – 5.0 micrometres
The cell's tail-like flagellum enables it to move. The cell wall is lined with frondlike pili.

3: **LUCA**
Last Universal Common Ancestor
The cell that links all life on Earth

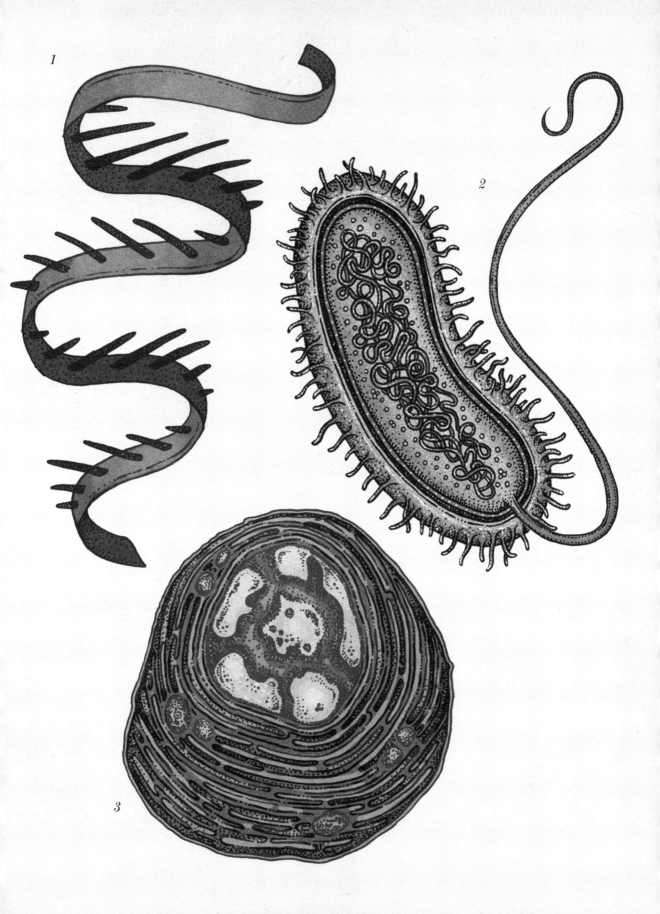

> PRECAMBRIAN
> 4.6 bya – 541 mya

Cyanobacteria

Among the first simple organisms to evolve was a group of bacteria called Cyanobacteria (blue-green algae). Made up of single prokaryotic cells, they produced oxygen using photosynthesis – a process by which some organisms can use the energy in sunlight to turn carbon dioxide gas and water into food. Oxygen gradually built up in the Earth's atmosphere, paving the way for new forms of oxygen-breathing life. The oldest known fossils were formed by layers of sediment which became trapped between mats of cyanobacteria. These fossils, called stromatolites, allow us to date cyanobacteria back at least 2.1 billion years.

Next to evolve were more complex types of cells, called eukaryotes. These are distinguished by their internal organs, or organelles, including the nucleus – a control centre for the cell that carries all of its genetic information.

More complex types of algae also started to appear. The most simple were single-celled organisms similar to cyanobacteria, but in some varieties, cells could link to each other end-to-end to form long algal strands. These were one of the first basic types of multi-cellular organism.

―――――――――――――― *Key to plate* ――――――――――――――

1: **Cyanobacteria**
Length: 0.5–1 micrometre
These photosynthesising bacteria first appeared around 2.1 billion years ago.

2: **Eukaryote acritarchs**
Length: 15–80 micrometres
A cell with a nucleus (eukaryote)

3 & 4: **Algae and algal strand**
Length: Approx. 1mm

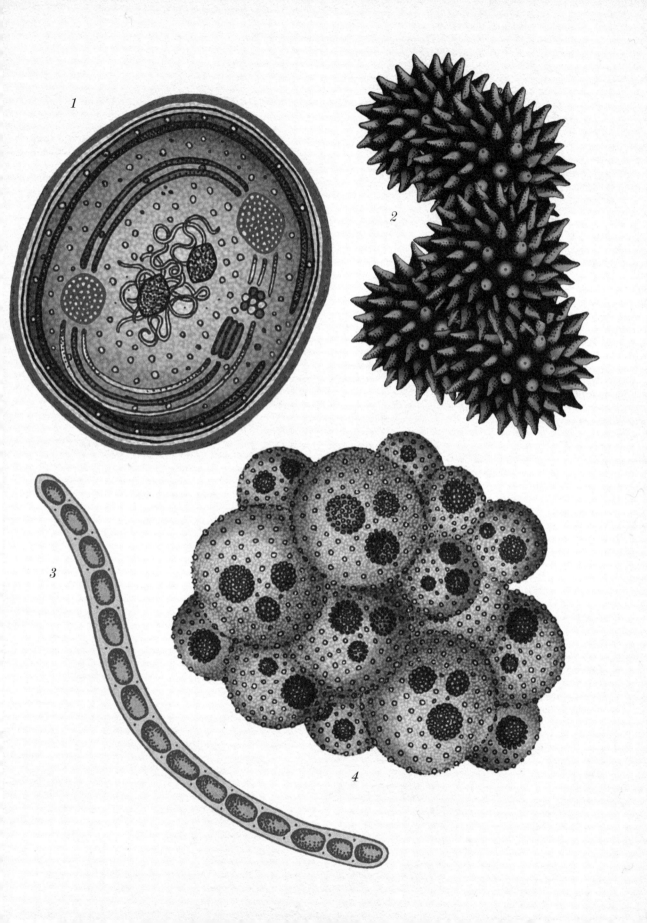

> PRECAMBRIAN
> 4.6 bya – 541 mya

Multicellular Animals

Gradually, early life forms developed into something more complex: the ancestors of all modern plants, fungi and animals. The first real evidence of this was discovered in the fossils of the Ediacara Hills of South Australia. They gave scientists their first look at the multicellular animals, such as *Spriggina* and *Funisia,* which populated the Precambrian seas.

These Precambrian lifeforms were mainly frond-shaped, wormlike, soft-bodied creatures. They all lived in the oceans, and many were attached to the seabed. Some, such as *Spriggina*, are thought to have been predators. This species is also one of the first to have had a distinct front and back end that served different functions. There is much debate about how to classify *Spriggina* and its fellow organisms, but it is widely believed to be the ancestor of the first arthropods – animals such as insects, with jointed bodies and an external skeleton.

Key to plate

1: Choanoflagellate
Length: 3–10 micrometres
Cells with a whiplike flagellum and collar

2: Charnia
Length: Up to 2m
Charnia grew on the deep seabed

3: Vaveliksia
Length: Up to 8cm
A spongelike organism

4: Funisia
Length: Up to 30cm
A wormlike organism

5: Spriggina
Length: Up to 5cm
A predatory organism with a fused head

6: Dickinsonia
Length: Up to 1m
A ribbed oval organism

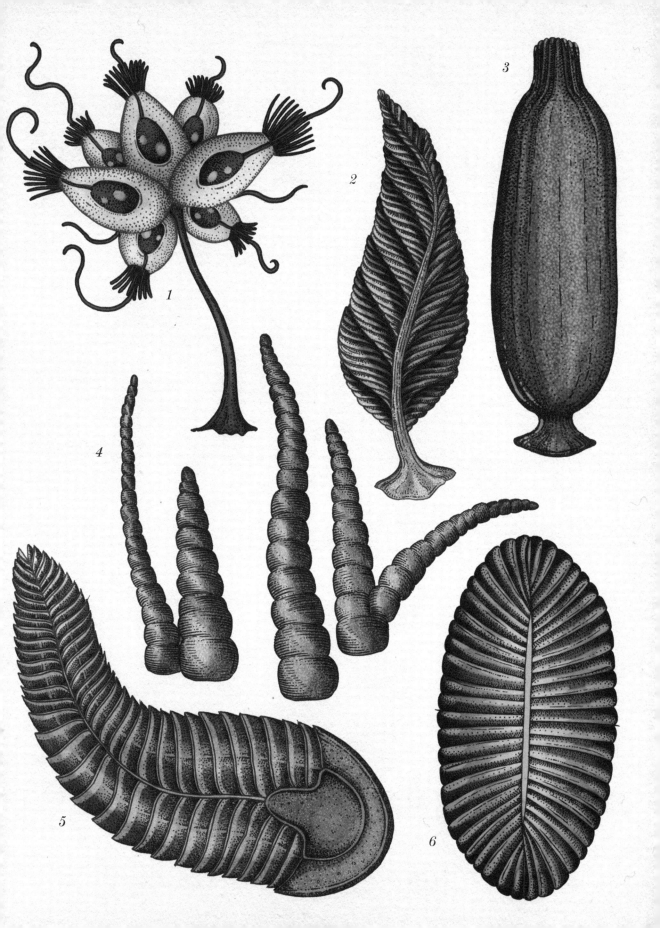

THE STORY OF LIFE

Gallery 2

Palaeozoic Era

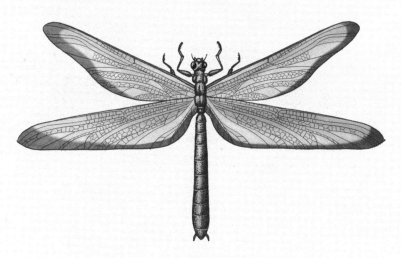

Cambrian Explosion; Cambrian Hunters; Ordovician Period; Silurian Period; Devonian Period; Acanthostega; Woody Plants; Carboniferous Period; Tetrapods; Permian Swamps; Sauropsids and Synapsids; Mammal-like Reptiles

CAMBRIAN
541 mya – 485 mya

Cambrian Explosion

The Cambrian Period saw an explosion in the diversity of life on Earth and the arrival of many of the major animal groups alive today. Lasting 20–25 million years, it was one of the most rapid periods of change in the planet's history. What sparked the incredible burst of new life in the Cambrian Period is not clear. Perhaps there was more oxygen in the atmosphere, or perhaps a warming climate had something to do with it.

The majority of the Cambrian animals, such as molluscs, worms and sponges, lived in shallow seas around the coast. They had much more ability to move than the organisms that came before them. This is because the Cambrian Period saw the evolution of the first chordates – animals with backbones – most likely resembling hagfish or lampreys.

The very first known chordate, *Pikaia*, probably swam over the seabed in an eel-like fashion, using its tail as a fin. At around the same time, the conodont also appeared. This long creature is known for its tiny toothlike fossils, which have been discovered in sites around the world. It is supposed they may have acted as teeth or been used to filter food from the water.

--- *Key to plate* ---

1: Dinomischus
Height: 2cm
Animal attached to seabed

2: Conodont
Length: 40cm
Eel-like early chordate

3: Pikaia
Length: 5cm
Early chordate relative

4: Crumillospongia
Length: 2.4cm
A sponge

Crumillospongia comes from the Latin words *crumilla*, meaning 'money purse' and *spongia*, meaning 'sponge'. This is on account of its saclike shape.

> CAMBRIAN
> 541 mya – 485 mya

Cambrian Hunters

Many of the new arrivals were arthropods – the ancestors of insects, spiders and crustaceans. Their hard bodies offered them both a defence against other creatures and a framework for supporting a larger body. They left their mark with a huge number of fossils, many of which were found at the Burgess Shale formation in British Columbia, Canada. These almost perfectly preserved fossils contain not only shells and teeth, but also muscles, gills and digestive systems, giving important clues to how these animals lived.

A group of the new arthropods called trilobites were the first animals to have complex eyes. They were preyed on by *Anomalocaris*, one of the most fearsome hunters in the Cambrian seas. But as hunters evolved, so too did their victims. *Wiwaxia* lived on the seabed and developed scales and spikes to ward off predators. Similarly, prickly *Hallucigenia* stood on seven pairs of legs and had two rows of spikes on its back. It was totally blind and relied entirely on its spikes for protection.

--- *Key to plate* ---

1: Anomalocaris
Length: Up to 2m
Largest Cambrian predator

2: Hallucigenia
Length: 3.5cm
Tubular organism with legs and spikes

3: Opabinia regalis
Length: Up to 7.5cm
Relative of the arthropods

4: Triarthrus
Length: 12cm
A trilobite – an arthropod with a three-lobed body

5: Wiwaxia
Length: 5cm
Sluglike creature with scales and spikes

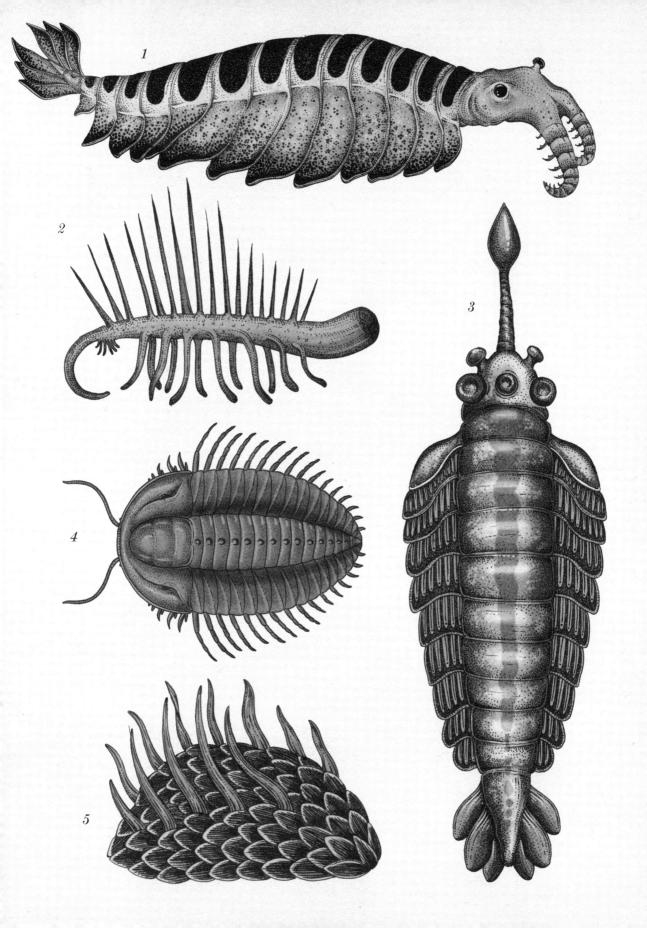

ORDOVICIAN
485 mya – 444 mya

Ordovician Period

During the Ordovician Period a rich variety of marine life flourished in the shallow seas. The Earth's climate was warm and wet, meaning conditions were just right for new kinds of life. There were squidlike nautiloids, the ancestors of cuttlefish and octopuses, which had gas-filled cavities in their shells to help them float and grasping tentacles for finding food. There was also a giant sea scorpion called *Pterygotus*, which was a distant relative of the modern king crab. It could have scuttled across the seabed or swum using its flattened tail, grabbing prey in its fierce pincers.

Fish, such as *Sacabambaspis*, became more widespread. Their bodies were protected by a bony shield and their mouths were jawless so food had to be strained from the water. In the oceans, large coral-like structures were home to sea lilies, which were anchored to the seabed. They collected food particles with feathery arms that waved in the currents. At the same time, arthropods moved towards shallow freshwater lagoons, and there is also evidence that the first simple plant-type organisms had begun to appear on land.

Key to plate

1: Orthoceras
Length: Up to 45cm
Early cephalopod, a group of animals which includes modern-day octopuses and cuttlefish

2: Sinuites
Length: Up to 2cm
Ordovician mollusc

3: Pterygotus
Length: 2m
Sea scorpion

4: Sacabambaspis
Length: 25cm
Early jawless fish

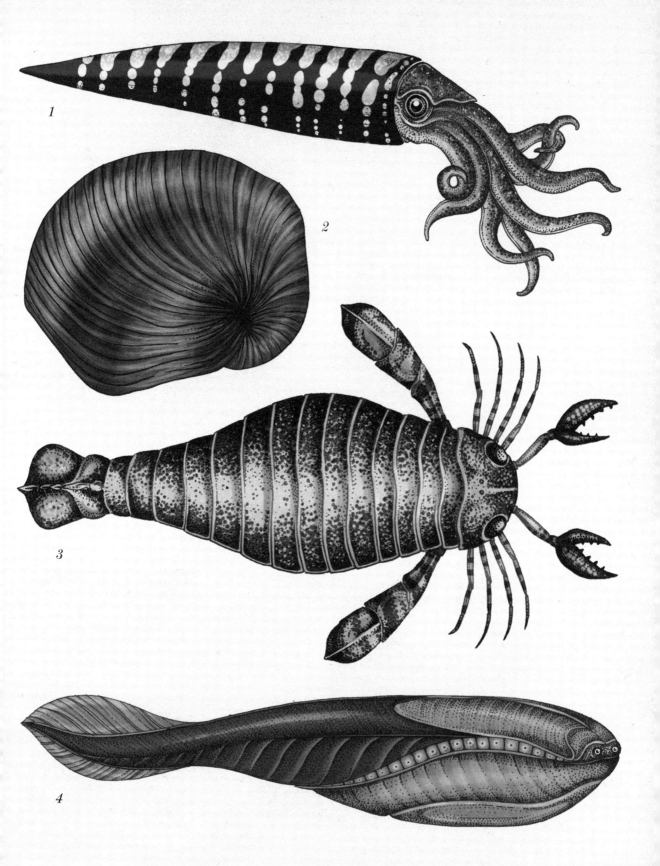

SILURIAN
444 mya – 419 mya

Silurian Period: First Land Plants

At the end of the Ordovician, freezing conditions gripped the Earth. Massive ice sheets formed and sea levels fell, draining shallow inland seas. At least half of all this extraordinary new life was completely wiped out.

During the Silurian Period that followed, the climate on Earth changed dramatically. Temperatures increased, resulting in melting ice sheets and rising sea levels. Giant reefs of coral, such as the now extinct *Halysites*, began to form and the first known freshwater fish appeared, as did the first jawed bony fish like *Acanthodes*.

Clear evidence that life had moved on to the land can also be seen in the fossil record. Early plants, such as *Cooksonia* and *Rhynia*, had small, stiff stems and branches, allowing them to stand upright. Like modern-day plants, they had vascular tissues, which are used to transport water and nutrients around their system. However, they had not yet evolved leaves. These primitive plants grew along the water's edge in coastal lowlands, creating vast wetland habitats – habitats that would later encourage animals to leave the water and walk on to the land.

--- Key to plate ---

1: Acanthodes
Length: Up to 30cm
Early jawed fish

2: Halysites
Length: 12cm
Coral

3: Cooksonia
Height: 6.5cm
Primitive land plant

4: Rhynia **stem and sporongium**
Height: 18cm

Primitive land plant
A sporongium is the part in some primitive plants where asexual spores are produced.

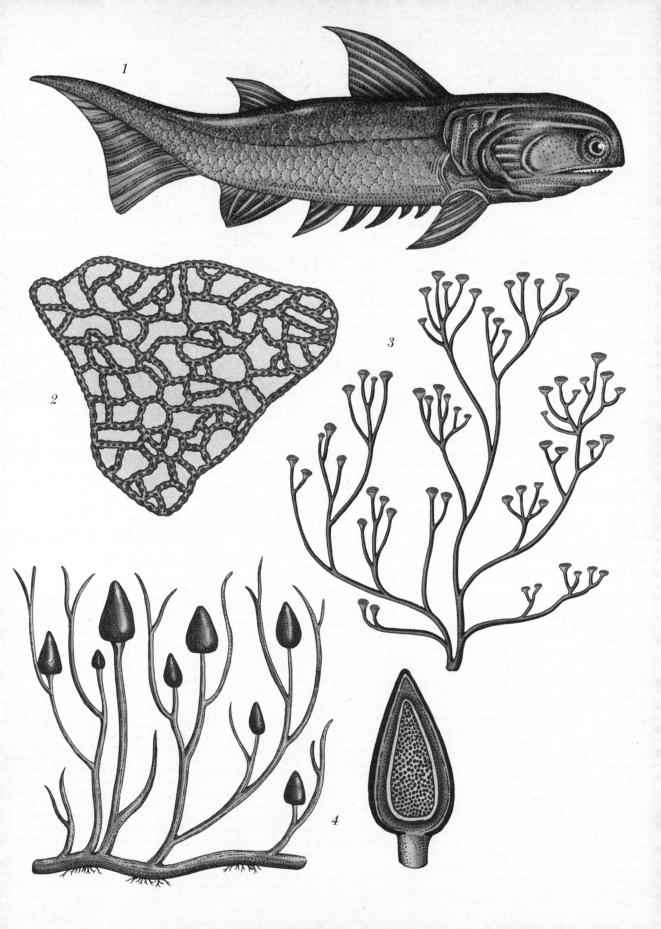

> DEVONIAN
> 419 mya – 359 mya

Devonian Period: the Age of Fish

The Devonian Period is also known as the Age of Fish, as several important species developed at this time. Among the biggest predators were armoured placoderms, such as *Dunkleosteus*, with powerful jaws and bladelike plates that acted as teeth. Some species grew up to 10 metres long and they were the dominant predators of the Devonian seas.

By this time, two other major groups of fish had become established: the bony fish and the cartilaginous fish. The cartilaginous fish had skeletons made of cartilage instead of hard bone. These ancestors of modern sharks and rays included the distinctive *Stethacanthus*, an early shark with an iron-shaped dorsal fin and a cluster of spiky 'scales' on its head. The bony fish were themselves split into two types; the ray-finned fish and the lobe-finned fish. The ray-finned fish, named after the thin bones supporting their fins, gave rise to most modern-day fish. Meanwhile, the lobe-fins, named after the thick, fleshy base to their fins, are thought to be the ancestors of all tetrapods – four-limbed land vertebrates, including dinosaurs, reptiles, amphibians, birds and mammals.

--- *Key to plate* ---

1: Dunkleosteus
Length: 10m
A placoderm

2: Eusthenopteron
Length: 1.8m

Lobe-finned fish and close to the ancestor of land animals

3: Cheirolepis
Length: Up to 50cm

Primitive ray-finned fish

4: Stethacanthus
Length: 70cm
An early shark with an iron-shaped dorsal fin

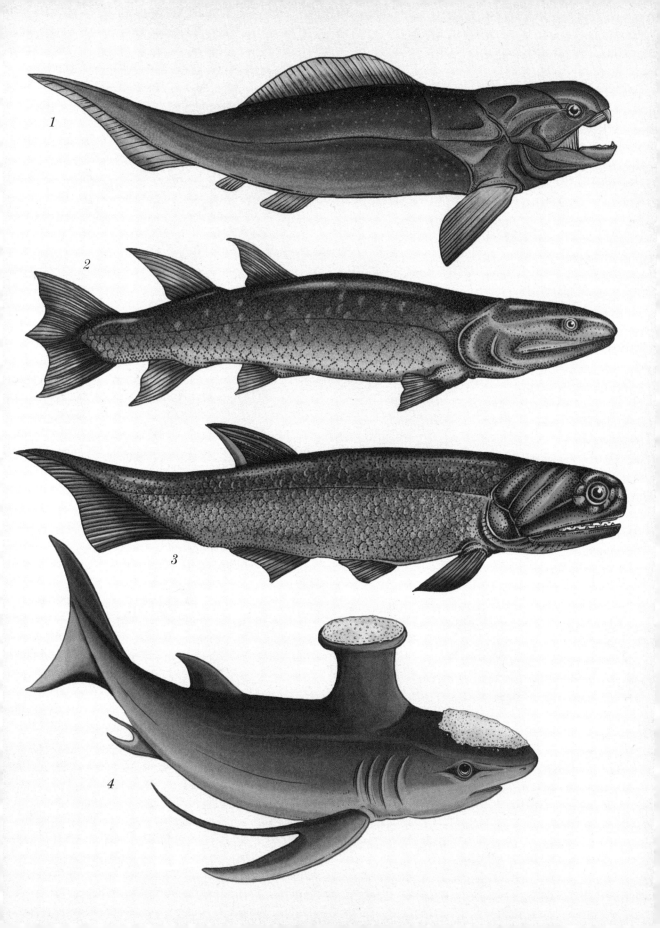

> DEVONIAN
> 419 mya – 359 mya

Acanthostega

Acanthostega, which appeared in the late Devonian, is an interesting example of the transition from lobe-finned fish to the first land tetrapods. It still had gills and a ray-finned tail, like a fish, but it also had small lungs and four limbs, each with eight digits.

1

The *Acanthostega* fossils found to date are thought to be those of juveniles. This raises the possibility that adults lived on land, in the same way that adult newts and frogs live on land while their young live in the water.

--- *Key to plate* ---

1: Acanthostega
Length: 60cm
Primitive tetrapod

Acanthostega is considered to be a transitional organism between water-bound fish and the first four-legged land animals.

> DEVONIAN
> 419 mya – 359 mya

Woody Plants

During the Devonian, plants were also developing at a rapid rate. By now they had many of the features we know in plants today – primarily a woody body that could support raised branches and leaf-type growths. In essence, they were the world's first trees. However, most plants did not yet have roots. Instead, species such as *Asteroxylon* had stems, called rhizomes, that grew horizontally underground and put out occasional upward-growing shoots.

The Devonian is also the time when the first plants with seeds evolved. The earliest seed plant known to us is *Elkinsia,* an early seed fern whose seeds grew right along its branches. Other plants, including horsetails, reproduced by releasing spores into the air from cones at the end of their branches. Some of these plants still survive today. Like their ancestors, modern-day horsetails favour wet, swampy ground.

Over time, the Devonian trees grew into massive forests, stretching along boggy coastal areas and growing up to 30 metres tall. These would provide the perfect environment for the diverse life forms that were making their way out of the seas and on to the land.

---------- *Key to plate* ----------

1: Elkinsia
Height: 50cm
Earliest known seed plant

2: Asteroxylon
Height: 40cm
Devonian land plant

3: **Horsetail and horsetail cone**
Height: 0.3–2m; cone: 2–6cm
Primitive treelike plant

CARBONIFEROUS
359 mya – 299 mya

Carboniferous Period: the Age of Insects

At the beginning of the Carboniferous Period, the climate was tropical and humid. Trees evolved roots, enabling them to grow even taller, and swathes of swampy forest covered the land, with giant horsetails, mosses and the first flowering trees. This dense mass of trees was eventually preserved in huge coal beds, which give the period its name, meaning 'carbon-bearing'.

The oxygen-rich air produced by the forests enabled insects and other arthropods to grow to enormous sizes – and with no reptiles or birds to prey on them, they flourished. *Arthropleura* was a two-metre-long poisonous millipede that probably fed on plants but may also have eaten smaller insects. It scuttled across the forest floor alongside gigantic cockroaches and spider-like creatures such as *Eophrynus prestvicii*.

It was also at this time that some insects grew wings and developed the power of flight. *Meganeura,* a huge dragonfly-like insect, had a wingspan of around 75 centimetres. Like dragonflies today, it would have hunted other insects in the air.

―――――――――――――― *Key to plate* ――――――――――――――

1: Meganeura
Wingspan: 75cm
Giant winged invertebrate

2: Arthropleura
Length: 2m
Giant land-dwelling invertebrate

3: Eophrynus prestvicii
Length: 2.5cm
Early relative of spiders

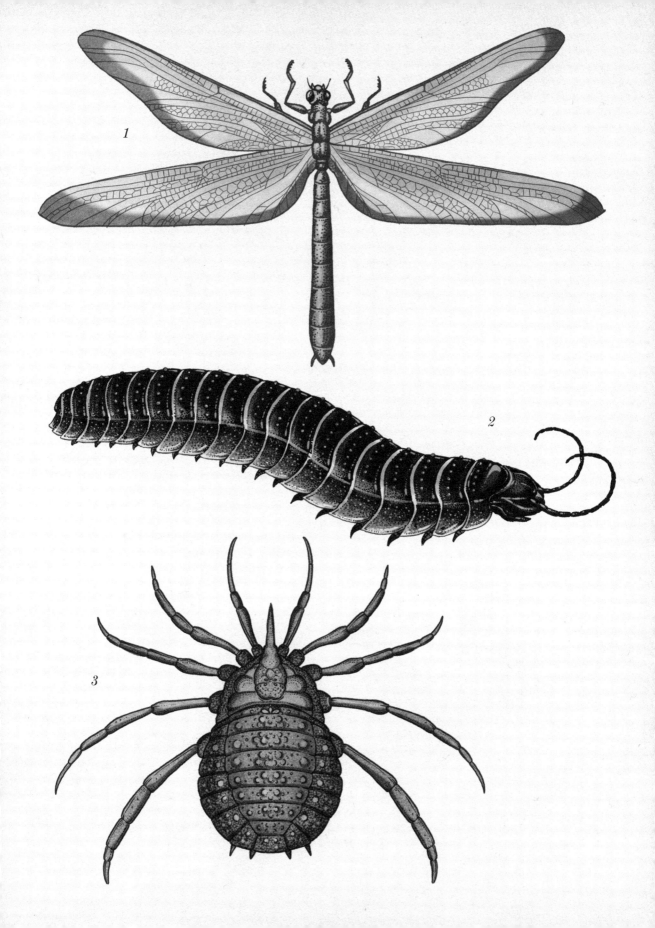

> CARBONIFEROUS
> 359 mya – 299 mya

1

Tetrapods

It was during the Carboniferous that tetrapods – animals with four legs – began to walk on land and evolve into amphibians. Some were like crocodiles with vicious teeth. Some even developed scaly skin, which stopped them drying out away from water.

Eryops was one such early amphibian, with a heavy body that grew up to three metres long. This lumbering creature hunted both in water and on land, but had to lay its soft-shelled eggs in the water, just as frogs and toads do today. It would have been well-suited to the swampy habitats of the time.

The first reptiles, which were small, scuttling animals, appeared towards the end of the Carboniferous Period. Unlike amphibians, they were able to lay hard-shelled eggs on land. The earliest reptile discovered so far is the 30-centimetre-long *Hylonomus*. This lizard-like animal probably lived on a diet of insects and millipedes.

---------- *Key to plate* ----------

1: Hylonomus
Length: 20cm
Earliest known reptile

2: Eryops
Length: 3m
Early amphibian

Eryops means 'drawn-out face' on account of the species' long snout shape.

> PERMIAN
> 299 mya – 252 mya

Permian Swamps

By the Permian Period, the swampy forests of the tropics were beginning to dry out, making conditions harder for the plants of spore-scattering species, which depend on wet conditions to survive. Many species that had thrived in the Carboniferous became extinct, and were replaced by early gymnosperms – cycads, conifers and other seed-bearing plants. Cycads evolved from more primitive seed-bearing plants like *Medullosa*, which, despite its fernlike shape, grew large seeds on individual leaves.

In time, more advanced plants such as *Glossopteris* evolved. Like modern-day conifers, it grew seeds on cones between its stems. *Glossopteris* reached up to 30 metres in height and dominated the environments of the time, growing in huge swathes right across the Southern Hemisphere. Towards the end of the Permian, the climate became drier and seed ferns like *Neuropteris* surpassed *Glossopteris* as the dominant plant type. Despite this, many modern conifer families, such as cypress and pine trees, can be traced back to the Permian, and conifers have thrived right up until today.

--- *Key to plate* ---

1: Medullosa
Height: 5m
Early seed plant

2: Neuropteris
Height: 10m
Widespread seed fern

3: Medullosa seed
Length: Up to 6cm
Cross-section of a seed from *Medullosa*, an early seed-bearing plant

4 & 5: Glossopteris leaves & plant
Height: Up to 30m
Widespread tree-sized seed plant. It was the dominant plant type in the Permian Period.

PERMIAN
299 mya – 252 mya

Sauropsids and Synapsids

On land, two groups of animals, which first appeared in the Carboniferous, dominated the landscape: sauropsids and synapsids. Synapsids are thought to include the ancestors of today's mammals, and sauropsids are thought to include the ancestors of reptiles (including dinosaurs) and birds.

Being cold-blooded, these reptile-type creatures had to find ways to deal with the Earth's variations in temperature, from below freezing at night to over 37°C during the day. Some, such as *Dimetrodon*, had sail-like structures on their backs to help regulate their temperature. The sails soaked up the sunshine during the day and the heat kept them warm at night.

--- *Key to plate* ---

1: Pareiasaurus
Length: 2.5m
Large plant-eating sauropsid – an early reptile

2: Dimetrodon
Length: 4.5m
An early synapsid, the group including the mammal-like reptiles and the ancestors of mammals. Over a dozen species of *Dimetrodon* have been named since its first discovery in 1878.

PERMIAN
299 mya – 252 mya

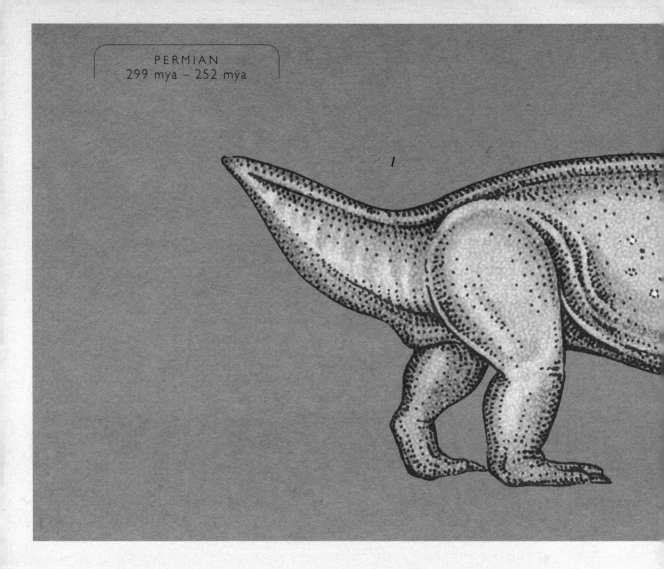

Mammal-like Reptiles

Synapsid reptiles – distinguished by the hole above their jaws – are often called 'mammal-like' reptiles, because of their characteristics and lifestyles, but they included the reptiles from which all mammals would eventually evolve.

Diictodon was one such synapsid with a large head, a beak and a pair of tusks. Five sharp claws on each hand and developed muscles made *Diictodon* an expert digger; a talent that was possibly useful when looking for somewhere to cool down during the scorching hot days. In fact it is thought that *Diictodon* had a lifestyle similar to that of gophers today.

It is also the oldest known example of sexual dimorphism – where males and females have different characteristics beyond their reproductive organs. Only the males had tusks, a trait that is seen in modern hoofed mammals, which often have tusks, horns or antlers for display or defence.

---────────────────────────── *Key to plate* ──────────────────────────────

1: Diictodon
Length: 45cm
Mammal-like synapsid
This species was one of the most successful synapsids of its time. The tusks on this individual *Diictodon* show that it was a male.

THE STORY OF LIFE

Gallery 3

Mesozoic Era

*The Triassic;
The Jurassic;
Archaeopteryx;
Cretaceous Period;
Tyrannosaurus rex*

TRIASSIC
252 mya – 200 mya

The Triassic

The end of the Permian saw the worst mass extinction in Earth's history. The reason behind it is not totally understood; it could have been caused by volcanic eruptions blocking out the Sun, by earthquakes or even by an asteroid crashing into Earth. Whatever the cause, it wiped out 90 per cent of all species on the planet.

Those that survived evolved to fill the empty Earth. The oceans teemed with ammonites, molluscs and sea urchins, and the first corals of the kind living today appeared. Reptiles including *Thrinaxodon* hunted on dry land, while the turtle-like *Henodus* plucked shellfish from the sea with its beak. Ancestors of creatures familiar today, such as frogs, crocodiles and snakes populated the freshwaters, and pterosaurs, a group of flying reptiles, filled the skies.

The biggest change in this period came towards its end, with the evolution of the first dinosaurs. These were all meat-eating theropods, mostly small and running on two powerful back legs. The first plant-eating dinosaurs evolved from these early meat-eaters. The Triassic Period then ended as it had begun, with another mass extinction, although the dinosaurs survived and went on to dominate the Jurassic Period which followed.

--- *Key to plate* ---

1: Sharovipteryx
Length: 25cm
An early gliding reptile, with a movement similar to that of modern-day flying squirrels

2: Henodus
Length: 1m
A turtle-like marine reptile with a broad shell and scaly skin

3: Thrinaxodon
Length: 50cm
A later synapsid. It may have survived the Permian extinction by burrowing.

JURASSIC
200 mya – 145 mya

The Jurassic

This was the age of the dinosaurs, but there was a lot more going on, too. A humid breezy climate encouraged palmlike trees, ferns and cycads with thick evergreen leaves to multiply. One such important plant, ginkgo, still thrives today. Without these Jurassic plants, the enormous meat-eating dinosaurs would not have survived, as they depended on herbivorous dinosaurs for food.

As well as giant dinosaurs such as the 30-metre-long *Brachiosaurus*, it is thought the first true mammals evolved during the Jurassic. *Megazostrodon*, for example, was a small furry animal a little like a shrew. In contrast to the huge dinosaurs, it measured just 10 centimetres long. Research has shown that the most developed parts of its brain were those that processed sounds and smells. This was probably in order for it to avoid being eaten by the roaming dinosaurs and other reptiles it lived alongside.

The warm seas of the Jurassic were filled with life of all kinds, from tiny plankton (still around today) to reptilian predators such as the dolphin-like ichthyosaurs and the long-necked, paddle-finned plesiosaurs.

─── *Key to plate* ───

1: Megazostrodon
Length: 10–12cm
Early mammal

2: Ginkgo
Height: Around 12m
Non-flowering seed plant

3: Archaefructus
Length: 50cm
One of the first flowering plants

4: Brachiosaurus
Length: 26m
Large plant-eating dinosaur

5: Elasmosaurus
Length: 14m
A large plesiosaur

> JURASSIC
> 200 mya – 145 mya

Archaeopteryx

Archaeopteryx was a small dinosaur, with wings and feathers like a bird but with other features more typical of predatory dinosaurs. About the size of a magpie, this species demonstrates the transition from feathered, predatory dinosaurs to true birds and is an important 'transitional fossil'.

Most *Archaeopteryx* fossils discovered so far have been found in the Solnhofen limestone beds in Bavaria, Germany, where the rock formation makes incredibly detailed fossils. Feather impressions discovered in the rock show that *Archaeopteryx* had feathers similar in appearance to flight feathers in modern birds. Together with the length and shape of the wing, and the arrangement of the feathers, this suggests *Archaeopteryx* could fly. While it was probably not airborne for long, it may have flapped short distances to chase after prey or escape danger.

However, unlike living birds, *Archaeopteryx* had a full set of teeth, claws on its wings, a 'killing claw' on its feet (similar to those seen on *Velociraptor*) and a long, bony tail. Altogether it had much more in common with meat-eating dinosaurs than with birds. Like other theropods, *Archaeopteryx* was a carnivore and would have hunted small prey, such as insects, small reptiles and amphibians. It probably snapped up prey in its toothed jaws but might have used its claws to snatch larger prey.

--- *Key to plate* ---

1: Archaeopteryx
Length: 90cm
A birdlike reptile

Archaeopteryx is commonly regarded as the transition fossil between feathered dinosaurs and the first true birds. Its name comes from the Ancient Greek meaning 'ancient feather'.

1

CRETACEOUS
145 mya – 65 mya

Cretaceous Period

During the Cretaceous, huge dinosaurs still dominated the land, but small mammals were also flourishing alongside them. Birds, too, were beginning to evolve, as ancestors to pelicans and cormorants filled the skies. The crow-sized *Confuciusornis*, from the early Cretaceous, is the first known bird to have had a beak.

Flowering plants really began to develop at this time, dramatically altering the look of the landscape. Beautiful species similar to modern magnolias covered the land along with the ancestors of *Ficus* and *Sassafras*, quickly outnumbering the ferns, cycads and other plants that had dominated the Jurassic. Their pollination – and therefore increased number – was helped by evolving insects like wasps, ants and beetles.

Throughout the Cretaceous, dinosaurs thrived and more specialised species evolved, all adapted to their lifestyle and habitat. Plant-eaters in particular became much more diverse, branching out into armoured ankylosaurs, spiky-backed stegosaurs, horned ceratopsians and thick-skulled pachycephalosaurians. New predators such as *Tyrannosaurus rex* also appeared.

──────────────── *Key to plate* ────────────────

1: **Cycad**
Height: Up to 4m
Non-flowering seed plant

2: **Cycad cone**
Length: Up to 0.75m

Cycads bear their reproductive organs – and seeds – in cones.

3: **Amborella**
Height: 8m
Early flowering plant

4: **Confuciusornis**
Length: 50cm
Early bird

> CRETACEOUS
> 145 mya – 65 mya

Tyrannosaurus rex

Tyrannosaurus rex belonged to a group of dinosaurs called the theropods, which was almost entirely made up of meat-eaters. Some theropods reached enormous sizes during the Cretaceous, and *Tyrannosaurus rex* was one of the largest ever to walk the Earth.

1

This huge predator had long legs, a powerful tail and two small arms. Its strong jaws and long serrated teeth were perfect for slicing through flesh. Like modern-day predators, *Tyrannosaurus rex* had forward-facing eyes, which would have given it excellent depth perception.

--- *Key to plate* ---

1: *Tyrannosaurus rex*
Length: 12m
Late Cretaceous predator

Tyrannosaurus rex whose name means 'tyrant lizard king' was one of the largest theropods, but was still smaller than the enormous meat-eaters *Giganotosaurus* or *Spinosaurus*.

THE STORY OF LIFE

Gallery 4

Cenozoic Era

*Palaeogene Period;
Ambulocetus; Cooling Earth;
Neogene Period;
Indricotherium; Megafauna;
Smilodon; Man*

PALAEOGENE
65 mya – 23 mya

Palaeogene Period

The end of the Cretaceous saw another mass extinction of life, widely believed to have been caused by a giant asteroid hitting the Earth. Dinosaurs and many other early creatures were wiped out. With the demise of the dinosaurs, some of the small, insectivorous mammals evolved into larger, carnivorous forms, such as *Sarkastodon*, feeding on the larger herbivores. Birds also flourished, with the giant, flightless *Gastornis* and others ruling the land, and with the appearance of the first known songbirds, including *Xenerodiops*.

The start of the Palaeogene is marked by the appearance of two new groups of mammals – the odd-toed ungulates, such as the horselike *Chalicotherium*, and the even-toed ungulates, such as deer – all plant-eaters that thrived on the wide grasslands of the Palaeogene. Early forms of many other animals began to appear too, including bats, bees, primates and rodents. *Halictus* is a bee genus that dates from this era and is still buzzing around today.

―――――――――――――― *Key to plate* ――――――――――――――

1: Sarkastodon
Length: 2.6m
Meat-eating mammal

2: Xenerodiops
Length: 50cm
Earliest known songbird

3: Chalicotherium
Length: 3m
Early plant-eating mammal

4: Halictus
Length: 10mm
Early bee

5: Hyracotherium
Length: 60cm
An early horselike mammal

> PALAEOGENE
> 65 mya – 23 mya

Ambulocetus

At the beginning of the Palaeogene, high temperatures led to tropical conditions right across the Northern Hemisphere, while warm oceans provided the perfect habitat for those reptiles that survived the Cretaceous extinction. Lizards, turtles and snakes clung to warm coastal waters, and there were even crocodiles in the Arctic.

An important fossil discovered from this time is *Ambulocetus*. This odd creature looked like a cross between a giant otter and a crocodile. It had a long,

tooth-lined snout and probably killed its prey as crocodiles do, dragging it under the water and holding it there. When scientists first discovered *Ambulocetus* they thought it could walk on land, like modern seals, which is why its name means 'walking whale'. However, recent research suggests this species spent most of its life in the water, just like whales today.

───────────────── *Key to plate* ─────────────────

1: Ambulocetus
Length: 3m
Early relative of whales

This species is thought to be the link between land animals and whales. As whales evolved over time, they lost their rear limbs, and their forelimbs evolved into flippers.

PALAEOGENE
65 mya – 23 mya

Cooling Earth

The Earth began to cool down towards the middle of the Palaeogene and deciduous trees overtook the tropical evergreen species. As the planet became still cooler and drier, the woodlands could not survive and grasses spread to more open areas. Grass has the advantage of growing back after being cropped and also contains silica, which was damaging to the teeth of many herbivores.

Only a few mammals, including some rodents, the ancestors of horses, and the ancestors of sheep and cattle, evolved special continuously-growing teeth in order to successfully feed on grass. These early ungulates, or hoofed animals, ranged in size from rodent-like animals to larger species that were around the size of a modern tapir. As ungulates grew in size and number, they gathered in groups for safety, and grazing herds became a typical sight of the Late Palaeogene.

Together with other flowering plants such as primitive daisies, the pollen and nectar produced by early grasses enabled insects to evolve alongside them. As insects flourished, giant ants, such as the five-centimetre-long *Titanomyrma*, scuttled across the plains.

Key to plate

1: **Titanomyrma**
Length: 5cm
Large ant

2: **Flowering grass head**
Length: Less than 2mm
Close-up of a grass head

3: **Flowering grass head**
Length: 2cm
Each grass head consists of many spikelets each made up of one or more flowers, protected by two scalelike glumes. The pollen-bearing anthers often protrude beyond the glumes when the grass is ripe.

4: **Grasses**
Height: Up to 2m

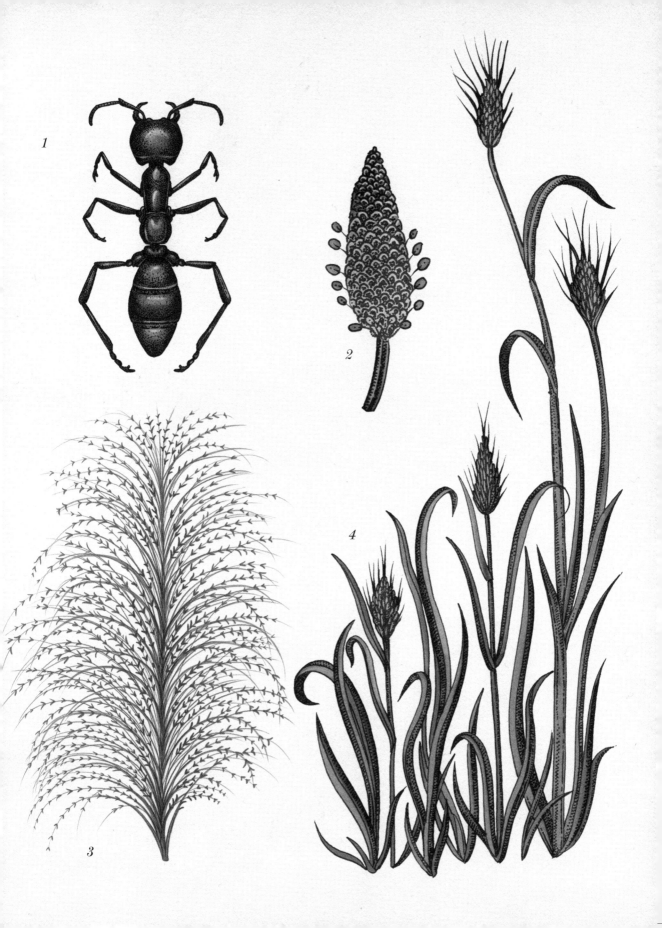

NEOGENE
23 mya – 2.58 mya

Neogene Period

The Neogene began with a series of ice ages as the Earth's climate cooled dramatically. Mountains rose up where tectonic plates pushed together, thick ice caps formed in the Arctic, and sea levels dropped, exposing land bridges between the continents – these enabled both fauna, and flora, such as wind-dispersed sycamore seeds, to spread across the globe. Plants also evolved to cope with the new conditions, with the arrival of new varieties, such as drought-resistant cacti. By the end of this period, 95 per cent of modern plants had appeared on the continents.

Grazing animals increased in number, while ducks, owls and cockatoos that would be recognisable today appeared. At the same time, primates were evolving and diversifying – around 100 species of ape lived during the Neogene, spread widely between Africa and Eurasia. The ancestors of humans were diverging from the ancestors of chimpanzees, each beginning their own evolutionary path.

─── Key to plate ───

1: **Monarch butterfly**
Wingspan: 10cm
A butterfly still seen today

2: **Musa**
Height: 4m
Red banana plant

3: **Sycamore fruit**
Width: 4cm
V-shaped 'wing' formation

4: **Cactus**
Height: Up to 20m
A prickly succulent

5: **Lily-like flower**
Height: Up to 1.5m
A plant with complex, often strangely-shaped, flowers

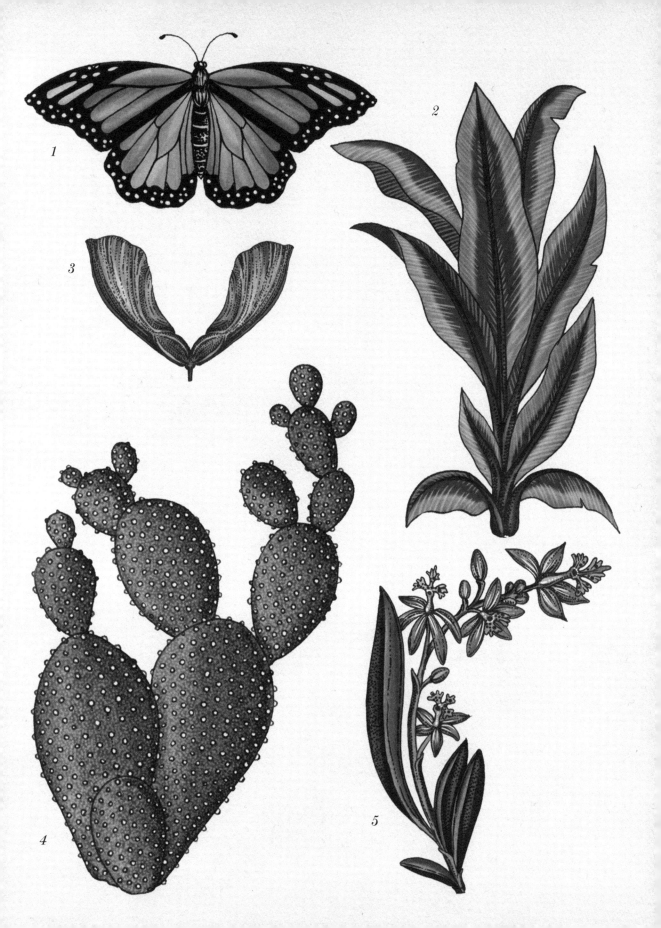

> NEOGENE
> 23 mya – 2.58 mya

Indricotherium

Mammals continued to evolve and multiply. By 2,500,000 years ago, many species had evolved into giants and huge creatures such as *Indricotherium*, a relative of the modern rhinoceros, stomped the Earth.

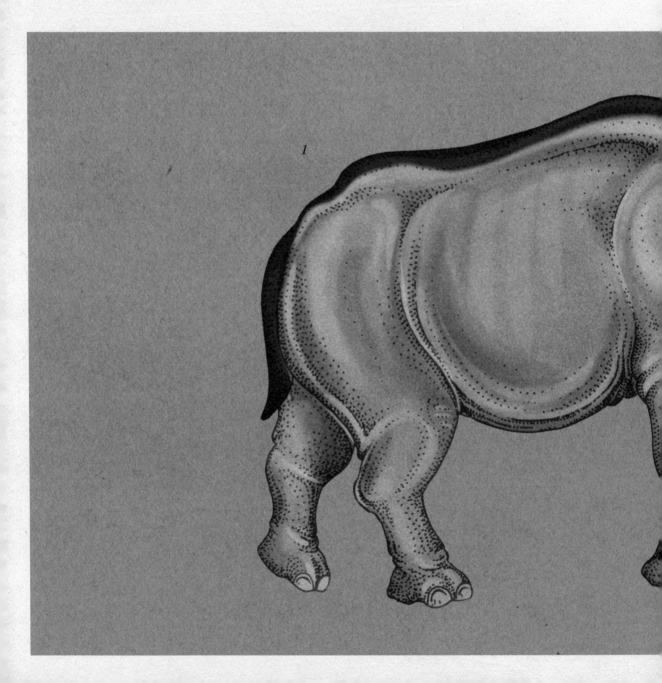

Standing five metres high and weighing around 30 tonnes – four times as heavy as a modern elephant – *Indricotherium* was the largest land mammal that ever existed.

--- *Key to plate* ---

1: Indricotherium
Height: 5m
Largest-ever land mammal

The enormous *Indricotherium* was so tall that, were it alive today, a modern-day human could have easily passed between its column-like legs.

NEOGENE
23 mya – 2.58 mya

Megafauna

The Neogene Period is known for its megafauna, a name that comes from the Greek word *mégalos*, meaning 'large', and the eighteenth century term *fauna*, meaning 'animals'. These giant mammals were many times larger than their modern-day counterparts – and it is believed they evolved to fill the gap left by the huge plant-eating dinosaurs. The resulting species were giants such as *Deinotherium*, a prehistoric relative of the modern-day elephant, with a larger, shorter trunk and downward-curving tusks. The mighty *Megaloceros* or 'Giant Elk' was another huge plant-eater, with antlers spanning nearly four metres. Of course, where there were plant-eaters there were also predators and, like their prey, they reached immense proportions. These ferocious beasts included the dire wolf, the sabre-toothed cats, and the giant hyena *Andrewsarchus*, whose skull was twice the length of a brown bear's.

However, around 12,000 years ago, the megafauna started to die off in what is known as the Quaternary extinction. This may have been due to climatic conditions as the planet cooled, a loss of food sources, or may have been partly due to the spread of the most dangerous predator of all – man.

Key to plate

1: Procoptodon
Height: 2m
Prehistoric kangaroo

2: Megaloceros
Height: 2m
Prehistoric deer

3: Glyptodon
Length: Up to 3.3m
A relative of the armadillo with a hard shell to protect it from predators

4: Deinotherium
Height: 4m
Huge prehistoric relative of modern-day elephants

> NEOGENE
> 23 mya – 2.58 mya

1

Smilodon

Other enormous creatures from this time included *Smilodon*, a sabre-toothed carnivore that prowled the Neogene woods and grasslands. So huge were its curved, serrated teeth, that they were always exposed, even when *Smilodon* closed its mouth.

With its huge bite, and strong, muscular legs, *Smilodon* would have leapt on its prey – quite possibly including early humans – from the high

branches of trees, then withdrawn to a safe distance to watch its dinner slowly bleed to death. This ambush technique, similar to that of modern jaguars, would have utilised the *Smilodon's* power, without risking damage to its relatively thin canine teeth.

Key to plate

1: Smilodon
Length: 2m
Prehistoric predatory cat

Smilodon is perhaps the best-known sabre-toothed cat of the Neogene. It is often mistakenly called the 'sabre-toothed tiger' despite the fact it is no relative of modern-day tigers.

> QUATERNARY
> 2.58 mya – present

Man

Humans belong to the primate family, which includes all monkeys and apes. Usually, humans and human-like species, properly known as hominins, are distinguished by their bipedalism (walking on two legs), their large brains and their ability to use tools.

The first primitive man was *Homo habilis*, who appeared around two and a half million years ago. We know from fossil finds that *Homo habilis* made and used tools, which suggests it may have scavenged large animals to eat, to supplement a diet of nuts, leaves, fruit and roots. *Homo habilis* wasn't tall compared to modern man, standing not much more than a metre high. Later *Homo habilis* lived at the same time as other human ancestors, such as *Homo erectus*, and was the first hominin to have body proportions similar to our own, with a long body and large brain.

Our closest extinct relatives were *Homo neanderthalensis*, or the Neanderthals, who evolved during the last Ice Age and lived side-by-side with modern man, or *Homo sapiens*. With their stocky bodies and wide noses for warming air, the Neanderthals were perfectly adapted for cold habitats. However, by 30,000 years ago, *Homo sapiens* were the only hominins left on the planet; the sole survivors of what had once been a diverse and sprawling family tree.

───────────── *Key to plate* ─────────────

1: Homo habilis
Height: 1.3m
Early hominin

2: Homo neanderthalensis
Height: 1.6m
Closest human relative,

commonly referred to
as Neanderthals

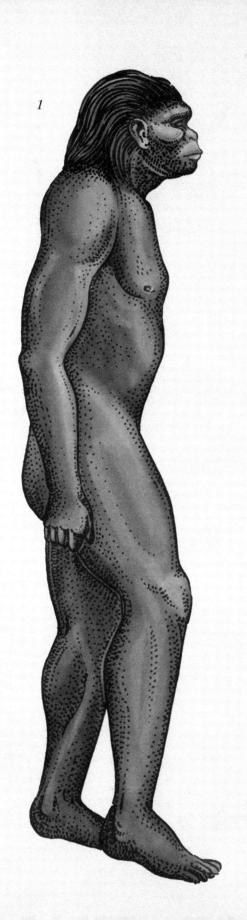

THE STORY OF LIFE

Library

Index

Index

Acanthodes 28
Acanthostega 32–33
algae 16
Amborella 54
Ambulocetus 62–63
ammonites 48
amphibians 30, 38, 52
Andrewsarchus 70
ankylosaurs 54
Anomalocaris 24
apes 66, 74
Archaefructus 50
Archaeopteryx 52
Arthropleura 36
arthropods 18, 24, 26, 36
Asteroxylon 34

Big Bang 7
bipedalism 74
birds 30, 42, 52, 54, 60
Brachiosaurus 50
Burgess Shale 24
butterflies 66

Cactus 66
Cambrian Explosion 22
Cambrian Period 10–11, 22–25
Carboniferous Period 10–11, 36–39, 42
Cenozoic Era 58–75
cephalopods 26
ceratopsians 54
Chalicotherium 60
Charnia 18
Cheirolepis 30
choanoflagellate 18
chordates 22

climate change 7, 22, 28, 64, 66, 70
Confuciusornis 54
conifers 40
conodonts 22
Cooksonia 28
corals 26, 28, 48
Cretaceous Period 10–11, 54–57
crocodiles 38, 48, 62, 63
Crumillospongia 22
cyanobacteria 16
cycads 40, 50, 54

Darwin, Charles 9
Deinotherium 70
Devonian Period 10–11, 30–35
Dickinsonia 18
Dicroidium 40
Diictodon 44–45
Dimetrodon 43
Dinomischus 22
dinosaurs 30, 42, 48, 50, 52, 54, 56–57, 60
Dunkleosteus 30

Earth
  atmosphere 16, 22
  formation 7, 14
Elasmosaurus 50
Elkinsia 34
Eophrynus prestvicii 36
Eryops 38, 39
eukaryotes 16
Eusthenopteron 30
evolution 9
extinctions, mass 48, 60, 70

ferns 34, 40, 50, 54
fish 26, 28, 30
  bony 30
  cartilaginous 30
  lobe-finned 30, 32
  ray-finned 30
flight 36, 48, 52
fossils 16, 18, 22, 24, 28, 33
Funisia 18

Gastornis 60
Giganotosaurus 57
ginkgo 50
Glossopteris 40
Glyptodon 70
grasses 64
gymnosperms 40

Halictus 60
Hallucigenia 24
Halysites 28
Henodus 48
hominins 74
Homo erectus 74
Homo habilis 74
Homo neanderthalensis 74
Homo sapiens 74
horsetails 34, 36
humans 7, 66, 74
Hylonomus 39
Hyracotherium 60

ice ages 28, 66, 74
ichthyosaurs 50
Indricotherium 68–69
insects 18, 24, 36, 52, 54, 60, 64
invertebrates 36

Jurassic Period 10–11, 48, 50–53

land, appearance of life on 26, 28
Last Universal Common Ancestor (LUCA) 14
life forms, earliest 14

mammals 30, 42, 50, 54, 60, 64, 68–69, 70
Medullosa 40
megafauna 70
Megaloceros 70
Meganeura 36
Megazostrodon 50
Mesozoic Era 46–57
molluscs 22, 48
multicellular animals 16, 18
Musa 66

natural selection 9
nautiloids 26
Neanderthals 74
Neogene Period 10–11, 66–73
Neuropteris 40

oceans and seas 14, 18, 22, 24, 26, 48, 50, 62
Opabinia regalis 24
Ordovician Period 10–11, 26–27
Orthoceras 26
oxygen 16, 22, 36

pachycephalosaurians 54
Palaeogene Period 10–11, 60–65
Palaeozoic Era 20–45
Pareiasaurus 43
Permian Period 10–11, 40–45, 48

photosynthesis 16
Pikaia 22
placoderms 30
plankton 50
plants 26, 28, 50, 66
    flowering 50, 54, 64, 66
    seed-bearing 34, 40, 50, 54
    woody 34
plesiosaurs 50
Precambrian Period 10–11, 12–19
primates 60, 66, 74
Procoptodon 70
prokaryotes 14, 16
pterosaurs 48
Pterygotus 26

Quaternary extinction 70
Quaternary Period 10–11, 74–75

reptiles 30, 39, 42, 43, 48, 50, 52, 62
    flying reptiles 48
    mammal-like reptiles 42, 43, 44–45
Rhynia 28
RNA 14
rodents 60, 64

sabre-toothed cats 70, 72–73
Sacabambaspis 26
Sarkastodon 60
sauropsids 42–43
sexual dimorphism 45
sharks 30
Sharovipteryx 48
Silurian Period 10–11, 28–29
single-celled organisms 14, 16
Sinuites 26
Smilodon 72–73

Spinosaurus 57
sponges 22
Spriggina 18
stegosaurs 54
Stethacanthus 30
stromatolites 16
synapsids 42–43, 44–45, 48

tetrapods 30, 32, 38–39
theropods 48, 52, 56
Thrinaxodon 48
timeline (life on Earth) 10–11
Titanomyrma 64
tool use 74
trees 34, 36, 40, 50, 64
Triarthrus 24
Triassic Period 10–11, 48–49
trilobites 24
Tyrannosaurus rex 54, 56–57

ungulates 60, 64

Vaveliksia 18
Velociraptor 52
vertebrates 30

Wallace, Alfred Russel 9
whales 63
Wiwaxia 24

Xenerodiops 60

Big Picture Press

Big Picture Press books are objects to be pored over and returned to, again and again... created by and made for the incurably curious.

Find us @BigPicturePress

BigPicturePress